LEAN A PART OF QUALITY MANAGEMENT

Copyright © 2024 Archana Sharm All rights reserved .ISBN

DEDICATION

We make a living by what we get but we make a life by give

Ti my family and friends my team mates,

Who have worked so long and so hard

To make my dreams to reality.

And People of the India and Abroad

Contents

ACKNOWLEDGMENTS iv

1. Genesis of Lean and Six Sigma 3
2. Genesis of Lean 17
3. Lean Principle 27
4. Lean Organization 37
5. Lean Waste 64
6. Tools And Technique- Kanban 72
7. Tools and Technique -Kaizen 88
8. Tools And Technique-Set Up Reduction 104
9. Tools And Techniques of Six Sigma 110
10. Tools and Techniques Poka Yoka 116
11. Tools and Technique Quality Tool 123
12. Tools And Techn.Job Standardization 129
13. Tools And Tech- Quality and Service 133
References 138
Glossary 140

ACKNOWLEDGMENTS

LEAN management concept is in the initial phase of implementation to Organisation. It is quite enthusiastic in implementing the Lean Process in Manufacturing and Service Industries to improve the productivity and total quality of products and services.

It requires dedication of full time Lean Resource Team has been constituted which has already started its operation. A group for Lean initiative has formed for guiding the activities of 3A and 5S of different capabilities have been formed in the achieves fruitful results in organisation.

LEAN brings out the basics of Lean Principles which will create general awareness amongst all people of organization and prove helpful yielding benefits of reduced cycle time letter worldwide.

GIVE A THOUGHT

"We are today Where yesterday's thought brought us and shall be tomorrow where today's thought carries us"

1.Genesis of Lean and Six Sigma

Six Sigma

Focus to Eliminate Defects and Process Variation:

Reduced Waste→Reduced Cycle Time Reduced→Variation Process Centering→Productivity Efficiency

Cost Benefits:

Specify Value→ Identify Value Steam Establish Flow →Create Pull Pursue Perfection→ Define Problem Measure As-Is Data Analyze Cause Improve Solution Control Process.

Lean Six Sigma Synergy (LSS)

Why Six Sigma and Lean need each other?

- Mutually Advantageous Conjunction
- LSS Combination is greater than Sum of Individual
- Mutually Advantageous Conjunction
- LSS Combination is greater than Sum of Individual

Six Sigma

Variation Focus Bottomline Oriented Synergies Between Lean and Six Sigma Fact Based Rigor Front Line Involvement Voice of Customer Systems Oriented.

Lean Six Sigma

How does Lean and Six Sigma complement each other?

Lean and Six Sigma: Lean to Exceptional Quality Lean and Six Sigma: Leads to Exceptional Quality

- LSS is about institutionalizing the cyclical process that eliminates waste and improves quality
- Success is in tightening the loop that reduces waste (Lean) and reduce variability Six Sigma (SS) while increasing the velocity with which value is created
- This not only reduces lead times and inventory, but also reduce variability and create more uniform output Less Waste.
- Improved Quality Less Variation Less Inventory Uniform Output Reduced Lead Time Synergies between Lean.

Application of Lean and Six Sigma Principles

Gather VoC→Define Problem and Goal→Define Y Metric→Collect Data→Evaluate MSA

Find Process Capability→Identify Root Causes→ Identify Solution→Plan Roll-out Implement Solution→Prepare Control Plan

Six Sigma What is Important? How are we doing? What is Wrong? What is to be done? How to Sustain?

DMAIC Lean

What is Important? As-Is Value Stream? How does it flow? How to improve flow? How to Optimize?

Specify What is Value to Customer (VoC)?

Map As-Is Value Stream→Measure Takt Time→VA, NVA

Analysis of As-Is Value: Stream Improve Flow by Eliminating Waste

Continuous →Improvements Through Kaizen

LSS→Define Y, CTQ,→Collect Data,→Map As-Is Value Stream,→Find Takt Time

Find Process→Capability→Value Analysis→Root Causes→Brainstorm→Solution→Eliminate Waste.

Future VSM→Improved Process→ControlPlan→Standardized Work, Kaizen.

DMAIC- CONTRO, IMPROVE, ANALYZE MEASURE, DEFINE

Six Sigma

1. Powerful Framework (DMAIC)
2. Statistical Tools
3. Visual Controls

The purpose of a Visual Factory or Visual Control is that the environment is enabled with distinct visual signals that hints or directs those in the environment about their next activity to be performed based on the signal. Visual Signals are delivered to inform, alert and motivate. It is a system of signs, information displays, layouts, color-coding, and Poka-Yoke or mistake proofing devices. Everyone involved can understand the status of the system just at a glance Enables the day-to-day workplace to be self-explaining, self-regulating, and self-controlling. Provides ability to manage variances in process and product at the Source while the product is being made Process conditions (machines, materials, manpower and methods) Product outcomes (Quality, Delivery, Cost)

4. Visual Controls Production Status Display Dashboard for Management Fault Message Display Error Indicators in MS Word Enable Visual Environment, Achieve Zero Defects
5. Uncover Root Causes
6. Understand & Reduce Variation
7. Do Things Right (Defect Free)

Lean Six Sigma Synergy

Application of Lean Tools and Techniques in DMAIC

Combination of Positives and Shortfalls

Six Sigma Shortfalls

↓ Issue Oriented

↓ Promote Belt Culture

↓ Rigid Process by Design

↓ No Prescriptive Solutions

Lean Positives

↑ Holistic, Systems oriented

↑ Front Line Involvement

↑ Standard Practices

↑ End State Driven Project Selection Lean Shortfalls

↓ Difficult to Understand

↓ Counter Intuitive

↓ Difficult to Scale

↓ Rigor Can Suffer

↓ Understanding Can Be Superficial

Six Sigma Positives

↑ Voice of the customer

↑ Scalable infrastructure

↑ Fact based rigor

↑ Bottom line oriented

↑ Top management relates easily

Synergies Between Lean and Six Sigma

Lean Project Guidelines:

The Future of Lean Six Sigma

The Ultimate Goal –> Transition from Push to Pull→Current State (Push) Future State→Steady State (Pull)

- Black Belt Initiatives – Projects assigned to Green Belts
- Issue Oriented - Reactive
- Goal Based Focus

- Improvement is Everyone's Job
- Engrained in Culture
- Do Project Whenever Required
- Stable Organization
- Green Belt Initiatives
- Project Improvements
- Value Stream Mapping
- Define Process Steps & Customers
- Measure Timing & Sequence
- Continuous Improvement Focus - Proactive

Lean Tool Sophistication Vs Time (Cultural Maturity)

Lean Usage Over Time

Identify Waste Scope (Expose Waste)

Simplify & Standardize (Reduce Variability)

Control (Stabilize Process)

- Understand the Processes
- Create Value Steam Mapping
- Analyze how output get to Customers
- Eliminate Non Value Added (NVA)Steps
- Minimize # of Ways to do Task
- Mistake Proof

- Jidoka: Stop if defect is detected

- Kanban: Information Flow

- Single Piece Flow : PDCA - Feedback Loop

Tool Complexity

- Pull : Customer Initiates Action
- Visual Management: Make processes visible for All
- FMEA: Risk Management
- Poka-Yoke: Mistake proof
- Kaizen, Takt Time, TPM, Level Load
- Verify each Step
- Shorten Feedback Loops
- Catch and Eliminate Defects at Source

Lean Project Maturity

Fundamental to Sophistication in Lean Projects

Selection of Lean and Six Sigma Projects

Guideline for Project Selection

Focus Area →Quickly Stretch →The Process thru Lean

Find Root Cause thru SS to shift mean →Reduce Process Variation & Waste using LSS→Control the End-to-End Process→VA high Value→Stream Mapping→Business Strategy

Voice Of Market

Customer→Business→Employee →Technology

Business Objectives

TopLevel→Dashboard(Indicators)→BigY's→Feedback→NV A high→Improvement→Opportunity Assessment Focus Area

2. Genesis of Lean

Genesis of Lean

History Timeline for Lean Manufacturing

1799-1850: Whitney's accomplishment apart from inventing Cotton Gin is his perfection of Interchangeable parts during manufacturing of Muskets for the U.S. Army at very low price Late 1890s: Time Study and Standardized work. Concept of Scientific Management Introduced by Taylor 1900s: Gilbreth Invented Process Charting / Flow Charting. Process Charts focused on all elements including Non-Value Add elements.

1910: Ford practiced the first comprehensive manufacturing strategy. People, Machines, Tooling and Products are arranged in continuous system for manufacturing.

1949 - 1975: Ohno and Shingo studied American, Ford and SQC practices of Ishikawa, Edwards Deming and Joseph Juran.

Developed Toyota Production System (TPS) and other techniques for World Class Manufacturing 1980 and Beyond: Application of TPS by various manufacturing companies.

James Womack's book "The Machine That Changed The World" brought in a new phrase-- "Lean Manufacturing" essentially, TPS. "Lean Thinking" published in 1996 by James Womack and Daniel

"Jone Womack and Daniel Jone"

Lean Manufacturing at Toyota

A Brief History 1902: Sakichi Toyoda, founder of Toyota Group invented an automated loom that stopped anytime a thread broke. This reduced defects and enabled operators to do multi-tasking. 1946: Post World War II, Toyota was manufacturer of small trucks and automobiles.

Findings after Benchmarked with Ford at Detroit: Productivity: 1/10th of Detroit, Production: Less than 1/100th of Ford, Japan Small and Diverse Market.

1950: Chief Engineer Taiichi Ohno Studied Henry Ford's writings, Visited Ford Plant. Begins "War on Waste".

1950 - 1960: Taiichi Ohno Studied Ford for years, created a manufacturing system with various techniques to reduce cost through elimination of waste

1980: IMVP Study identified "Lean Manufacturing"

1984: Toyota becomes North American Manufacturer with Joint Venture at NUMMI Plant in California

1990 and Beyond: Lean Production Strategies adopted in numerous service and manufacturing Industries IMVP: International Motor Vehicle Program NUMMI: New United Motor Manufacturing Inc: JV between GM and Toyota.

Definitions of Lean Lean (or One-piece-flow or JIT or)

"A philosophy that shortens the time line between the customer order and the shipment by eliminating waste (non-value-adding activities)." "The continuous movement of products and information through the value stream. The goal is to minimize 'idle time' which equates to waste . . ." (C. Fiore; Honeywell presentation to ASQ) "Lean is a business environment where waste is identified continuously and eliminated passionately "WINOC: Work in Northeast Ohio Council (October 18, 2005)

LEAN

It is a Japanese Culture where we used focus on waste elimination in life. It applies also in setup of Organization

Its use for Mass Production Firstly introduced by Toyota in Manufacturing Industry of Automobile.

Lean is Japanese word which used a process improvement strategy that seeks to eliminate inefficiencies in a company's process flow by identifying the causes of waste or redundancy and developing solutions to address them for least waste. Using different technique to get quality and clean space and hassle-free work at work space. It is best use in aspect of man material and machine.

It includes Kanban and Kaizen to analyses and control the wastes within time face to get Quality output of the products or services.

Kaizen is the philosophies of Daily Improvements and Good Thinking, Good Products, Toyota Production System (TPS) has evolved into a world-renowned production system. Even today, all of Toyota is implementing kaizen to TPS day and night to ensure its continued evolution

Toyota Motor Corporation's Toyota Production System

Toyota Production System (TPS) is a way of making things that have become known and studied worldwide.

It is based on the premise of making work easier for workers and People. The objective is to thoroughly eliminate waste and shorten lead times to deliver vehicles to customers quickly, at a low cost, and with high quality. This production system is pursued in all areas of Toyota Motor Corporation, including vehicles and services, and All employees implement daily incremental kaizen.

A production system based on the philosophy of achieving the complete elimination of waste in pursuit of the most efficient methods for short time. It is coming into KANBAN were FIFO, LIFO Pull method is associated.

LEAN Over traditional Draw backs

Unorganized

Untidy

Uncontrollable

Too much wastes

Long setup

Long Lead time

Demoralize

No cost-effective process.

LEAN Process is Waste elimination

Organized

Clean and Tide

Proactive and Controlled

Using Kanban Short and Setups Through

Kaizen Environment for Improvements

No Wastes

Short Leads Times

Improve Morale

Save Money and Cost

Toyota Production System -Foundation of Lean Manufacturing

Just in Time (JIT) "The right part at the right time in the right amount" Continuous Flow Pull Systems Level Production Built-in-Quality (Jidoka) "Stopping at Abnormality" Manual / Automatic Line Stop Mistake Proofing Visual Control Operational Stability Standardized Work Robust Products and Processes Total Productive Maintenance Supplier Involvement High Quality, Less Cost, On-Time Delivery through Shortening Production Flow by Eliminating Waste Culture Flexible, Capable, Highly Motivated People

Lean Fundamentals GE's House of Lean Enterprise Core Principles:–Start with customer focus–Continuous improvement–Employee participation–A new stage of Six Sigma Built on a Foundation of Six Sigma and Digitization "JIT"– Just In Time; Deliver what customers want, when they want it "Heijunka"– Sequencing and Level Loading "Jidoka"– Stop at every abnormality; Human intelligence built into machines.

LEAN of the Toyota Production System

The Toyota Production System (TPS), which is based on the philosophy of the complete elimination of waste in pursuit of the most efficient methods, has roots tracing back to the automatic loom invented by Sakichi Toyoda, the founder of the Toyota Group. TPS has evolved through many years of trial and error to improve efficiency based on the Just-in-Time concept developed by Kiichiro Toyoda, the founder of Toyota Motor Corporation.

At the root of this is Sakichi Toyoda's idea of "doing things for others." As he sought something he could do that would benefit the world, he focused on making things easier for his mother, who worked late into the night operating a manual loom. The automatic loom that he invented not only automated work that used to be performed by hand but also built the capability to make judgments into the machine itself. By eliminating both defective products and the associated wasteful practices, Sakichi succeeded in rapidly improving both productivity and work efficiency. This is where the concept of jidoka was born.

Kiichiro Toyoda advocated Just-in-Time based on this strong conviction:

"A complete car cannot be built if even one part is missing."
Coordinating this is no small task. However, without this control, we could have a mountain of parts and still not be able to build a car.

"No amount of money will suffice if we don't think of a unique way to organize these tens of thousands of parts."

Via the philosophies of Daily Improvements and Good Thinking, Good Products, TPS has evolved into a world-renowned production system. Even today, all of Toyota is implementing kaizen to TPS day and night to ensure its continued evolution.

We carry on TPS around the world with a strong

Lean Fundamentals
Toyota Production System - Foundation of Lean Manufacturing

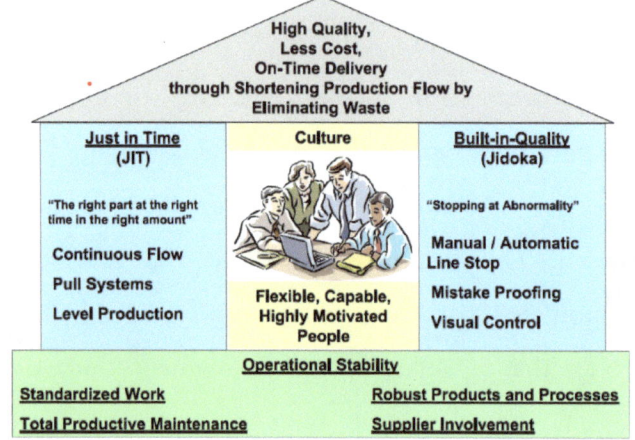

Lean Fundamentals
GE's House of Lean Enterprise

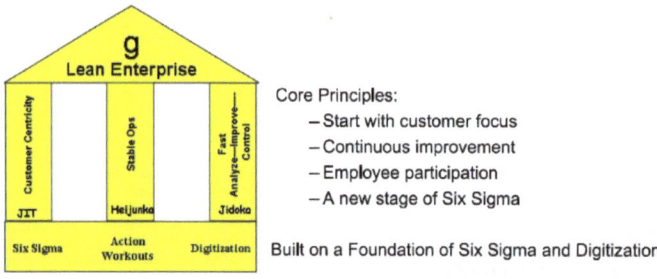

"**JIT**" – Just In Time; Deliver what customers want, when they want it
"**Heijunka**" – Sequencing and Level Loading
"**Jidoka**" – Stop at every abnormality; Human intelligence built into machines

5 LEAN FOCUSES

- Work for the customer. the primary goal of any change you want to implement should be to deliver maximum benefit to

the customer.
- Find your problem and focus on it. ...
- Remove variation and bottlenecks. ...
- Communicate clearly and train team members. ...
- Be flexible and responsive.

3. Lean Principle

LEAN PRINCIPLES

Specific Value   From Customer Perspective

Value Stream   Map Value Stream & Eliminate non-value-added activities

Flow without interruptions Let the value-added activities flow, Kanban (Pull)Produce as per customers demand Kaizen Pursue Perfection Be tireless in rooting to eliminating waste.

Principles of Lean

1. Specify Value – Why, What, How?
2. Why Value Customers?

Specify Value Identify Value Stream Businesses exists because of customers. Precise understanding of customer requirements and delivering at the right time. at an appropriate price, as defined in each case by the customer is the key to a Lean organization

3.What is Value?

Flow Perfection Pull Flow Elements of Service or Products in a form the customer is willing to pay for. It is defined by the customer and created by the producer

4.How is Value Determined?

Pull Specify Value Perfection What's Value in the eyes of the customer? Identify Value Stream Specific products or services are evaluated on which features add / create value. The evaluation is made from the internal and external customer standpoints. Value determination can be from the perspective of the ultimate customer or a subsequent process.

Examples of Value, specified for a Product or Service→ Market Profitability of the Product, Coverage of Operator Instruction User Manual, Number of orders bagged per month, etc.

Identify Value Stream

5. What is VA and NVA?

6. What is a Value Stream?

Identify Value Stream All of the actions both value adding and non-value adding, required to bring a product or service from order to delivery. It is a string of all activities required to process information from the customer and to transform the product on its way to the customer

7. What are Value Added Activities?

Any activity that transforms processes, materials or information The customer values that activity and willing to pay Done First Time Right

8. What is Non Value-Added Activities?

NVA Inevitable Flow Perfection Pull Flow- Pull Specify Value Perfection Identify Value Stream.

9. What are the chain of steps that create value?

Activities with no value but can't eliminate due to technology or equipment Regulatory, customer mandated, legal. Necessary because of risk tolerance, 'buffer' NVA Waste Activities that consume cost, time & resources but create no value in the eyes of the customer.

Identify Value Stream---VA and NVA

Analysis Value Stream Mapping (VSM) is the first step for an organization's war on waste.

VSM helps the organization understand how value added and non-value-added activities impact their products and processes as it progresses from raw material to handover to the customer. VSM results in identifying wastes in the operational

Identify Value Stream – VA Ratio

10. What is VA / NVA Analysis?

$$Value\ Added\ Ratio\ (VAR) = \frac{Value\ Added\ Work}{Total\ Work}$$

Value-Added (VA) Activities must meet the following three criteria: processes triggering elimination or reduction of the waste. Value-Added (VA) Activities must meet the following three criteria:

1. The customer must be willing to pay for the activity
2. The part or object must change
3. It must be done right the first time

Identify Value Stream – Value Stream Mapping

The Value Stream Map is used to illustrate both the 'Current State' and the desired 'Future State' of a process. A current-state map follows a product's path from order to delivery to determine the current conditions.

A future-state map shows the opportunities for improvement identified in the current-state map to achieve a higher level of performance at some future point.

11. How is a Value Stream Map used to validate improvement?

With multiple products and processes in an organization, the Product Family Matrix can be used to help identify the various Value Streams.

12. How is a Value Stream Identified for Lean Study? Example:

- A company having 5 brands (products) can be visualized as having 5 value streams
- Strategic Analysis done on all factors to understand how well each brand performs in the market with respect to profit, demand, price, features, support and services
- Results show that brand C has a good demand but is competitively disadvantaged in other factors
- This means that the correct Value Stream will be Brand C for applying Lean Study

13. What are the steps to perform Value Stream Analysis?

- 1. Identify the correct Value Stream for study
- 2. Create the "Current State" Value Stream Mapping
- 3. Identify and Diagnose Non-Value Added activities
- 4. Brainstorm Improvements and suggest changes
- 5. Create the planned "Future State" Value Stream Mapping
- 6. Create an action plan to implement changes to make it the future state VSM.

Toyota says companies don't get proficient at value stream mapping until they've done it at least seven times on the same process !!! – (Lean Manufacturing .com)"Key Tenets of "Lean"→ Identify NVA Activities and Find Ways to Eliminate Waste" Example 1: Information Technology Specify Value→Identify Value Stream→Flow Pull Perfection→Principles of Lean

Example 2: Manufacturing

Specify Value→Identify Value Stream→Flow Pull Perfection

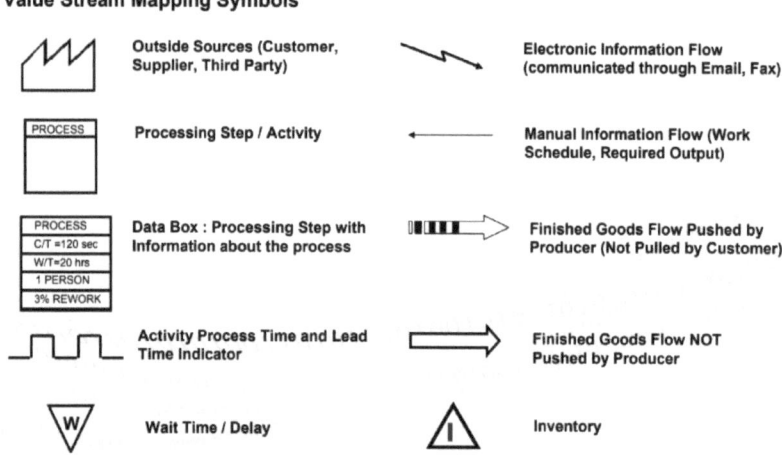

The rising success of Lean Manufacturing has been well-recognized in businesses for years now. Its core strategy for cultivating a culture of continuous improvement for business growth has been adopted by organizations from all sectors. With its techniques and tools to reduce waste and increase quality, the Lean manufacturing concept has been yielding notable results.

This practical guide covers the Lean manufacturing principles, how its history unfolded, the goals and benefits it can bring, and how to apply it in practice.

The idea of eliminating waste originates from the Toyota Production System. Taiichi Ohno, who is considered one of the founding fathers of Lean Manufacturing, dedicated his career to establishing a solid and efficient work process.

During his journey, Ohno described three major roadblocks that can influence a company's work processes negatively:

Muda (wasteful activities),

Muri (overburden),

Mura (unevenness).

Based on his observations and deep analysis, he categorized the 7 types of waste (7 Mudas), which later became a popular practice for cost reduction and optimizing resources. Lean Principles

Lean Management Approach is JIT (Just In Time) without waste with quality at competitive work for the customer. The primary goal of any change you want to implement should be to deliver maximum benefit to the customer. ...

Find your problem and focus on it. ...

Remove variation and bottlenecks. ...

Communicate clearly and train team members. ...

Be flexible and responsive.

LEAN Based on to reduce the Work based on 3A

AIM-------- LEAN Aim Efficient Management with Less Waste

ATTITUDE--- LEAN Attitude based on Zero Tolerance for Waste

ACTION------ LEAN –Action Continuous Elimination having Of Waste

LEAN principle used in eliminating the Waste/wasteful activities in all areas under any control and to bring in Continuous Improvement in achieve perfection include in action.

LEAN benefits

- Work Simplification
- Uninterrupted work
- Clean work area
- Good quality output
- Less fatigue
- Time saving
- Cost reduction
- Goals
- Identify value stream encompassing all business processes, continuously eliminate waste and impowering them to achieve:
    Best Quality On Time, Delivery,
    Inventory Deduction,
    Higher Capacity

4.Lean Organization

LEAN ORGANIZATION

Lean Organization is the identification and removal of waste so that everyone becomes more productive efficient result oriented and customer focused,

How we can achieve it

It is done by -------

- Carrying out of 5S (Sort, Set, Shine, Standardize. Self-discipline)
- Create Flow KANBAN (Produce only what is needed by the next person/or customer)
- Put in visual controls. (Make a chart to show the actual against the schedule.
- Job Standardizations (Draw up procedures and standards to ensure repeatability.
- Set up reduction (After completing a job, see if you can reduce the time taken before starting the new job.

Continuous improvement (Keep on applying steps1 to 5 in all that you do.

Benefits of LEAN

It helps to reduce cost, maintain quality & Stick to delivery.

"Being things more efficiently & customer focused should be a way of life"

ORIGIN OF THE TOYOTA PRODUCTION

A production system fine-tuned over generations

Roots of the Toyota Production System

Steps to Lean Organisation.

Cultural change Superficial to Deep

Phase 1 Lean Assessment evaluation analysis

Phase 2 Lean Preparation / education

Phase 3 Lean Pilot Site

Phase 4 Lean Mobilizes

Phase 5 Expansions to Supply Chain

Phase 6 Sustain

Toyota Production System

A production system based on the philosophy of achieving the complete elimination of waste in pursuit of the most efficient methods.

Toyota Motor Corporation's Toyota Production System (TPS) is a way of making things that have become known and studied worldwide.

1. Andon ('Sign' or 'Signal')

A visual aid which highlights where action is required (eg. the flashing light in manufacturing plants that indicate the line has been stopped by one of the operators due to some irregularity). Andon is a typical tool to apply the Jidoka principle (also referred to as 'autonomation'), which means the highlighting of a problem, as it occurs, in order to immediately introduce countermeasures to prevent re-occurrence.

Originating from the word for a paper lantern, it is a term that refers to an illuminated signal notifying others of a problem within the quality-control or production streams. Activation of the alert – usually by a pull-cord or button – automatically halts production so that a solution can be found. The warning lights are incorporated into an easily visible, overhead signboard, which also identifies the area or specific workstation that has the problem.

The frequency and nature of these occasional issues are analysed as part of Toyota's programme of continual improvement.

2. Gemba or Genba

(The place where the real work is done)

Now adapted in management terminology to mean the 'workplace' or the place where value is added. In manufacturing, it usually refers to the shop floor.

Gemba, or Genba (as it is also spelt), refers to the factory floor or manufacturing floor; often an open-plan environment where each individual's work and actions are visible to others.

The car is built exclusively at a new Toyota Gazoo Racing production facility created within the Motomachi plant in Japan. This is the most specialised manufacturing plant in Toyota's worldwide repertoire, famously home to the iconic Lexus LFA supercar and the facility where the hydrogen fuel cell Eye-catching design effortlessly blends with innovative new technology. The Mirai is the next step in the age of zero harmful tail pipe emissions, powered by electricity made in its innovative fuel cell stack that fuses oxygen and hydrogen together and flagship Lexus LC coupe are currently produced.

This visibility is exploited in order for third-parties – usually management or section leaders – to conduct regular *Gemba Walks* in order to identify areas where potential improvements might be made, and to better understand the workload of each associate. Walks around the frontline environment of the Genba also ensure that the production system is correctly adhered to.

3. Genchi Genbutsu

(Go and see for yourself)

The best practice is to go and see the location or process where the problem exists in order to solve the problem quickly and efficiently. To grasp problems, confirm the facts and analyse root causes.

Closely related to the need to walk the *Genba*, this key principle suggests that to truly understand a situation you need to visit in person. The Toyota Production System requires a high level of management presence on the factory floor, so that if a problem exists in this area It should be first of all correctly understood before being solved.

The nature of the phrase is less about the physical act of visiting a site but more to do with a personal understanding of the full implications of any action within an environment as a whole.

4. Hansei (Self-reflection)

Even if a task is completed successfully, Toyota recognises the need for a *hansei-kai*, or reflection meeting; a process that helps to identify failures experienced along the way and create clear plans for future efforts.

An inability to identify issues is usually seen as an indication that you did not stretch to meet or exceed expectations, that you were not sufficiently critical or objective in your analysis, or that you lack modesty and humility. Within the *Hansei* process, no problem is itself a problem.

5. Heijunka (Production smoothing)

A levelling technique to facilitate Just-In-Time (JIT) production and to smooth out production in all departments, as well as that of suppliers over a period of time.

A vital technique for reducing waste and improving production efficiency by levelling fluctuations in performance within the assembly line. Fluctuation normally occurs through either customer demand or within production itself.

The Toyota Production System uses *Heijunka* to solve the former by assembling a mix of models within each batch, and ensuring that there is an inventory of product proportional to the variability in demand. Furthermore, the disruption of production flow is minimised by making sure that components are sequenced to be available in the right quantity and at the right time, while changeover periods for vital processes such as die changes within the steel presses are as short as possible; often in as little as three minutes.

6. Jidoka (automation with human intelligence)

One of the main principles of the Toyota Production System, It is the principle of designing equipment to stop automatically and to detect and call attention to problems immediately, whenever they occur (mechanical *jidoka*).

In the Toyota Production System, operators are equipped with the means of stopping production flow whenever they note anything suspicious (human *jidoka*). *Jidoka* prevents waste that would result from producing a series of defective items.

It also liberates operators from controlling machines, leaving them free to concentrate on tasks that enable them to exercise skill and judgement, instead of over watching each machine continuously.

A cost-effective quality control process that combines automation with a human's ability to quickly detect abnormalities, interrupt production, and then correct them before resuming. Employing *Jidoka* principles throughout the production process is a vital element of the Toyota Production System, forcing imperfections to be immediately addressed by self-inspecting workers and thereby reducing the amount of work added to a defective product.

Some automated machines can also function in the

detection process, allowing human operatives to only be engaged when alerted to a problem. Full application of *Jidoka* means that the process which created any issue is subsequently evaluated to remove the possibility of re-occurrence.

7. Just-In-Time (JIT)

The Toyota Production System is dictated by the needs of the customer, as we don't produce anything until there is a need for it. Just-In-Time production means only making what is needed, when it is needed and, in the amount, needed. TPS operates a 'pull' system. When each vehicle is made to order, a signal is sent for parts to be replaced, thus maintaining the parts and materials inventory at a balanced level. Production and transport take place simultaneously throughout the production sequence.

JIT is a method that minimises the generation, purchasing or holding of component parts as stock items prior to full assembly line production. The primary objectives are to save warehouse space and unnecessary cost-carrying and to improve efficiency, which means organising the delivery of component parts to individual work stations just before they are physically required.

To apply this flow efficiently means relying on ordering signals from *Kanban* boards or by forecasting parts usage ahead of time, though this latter method requires production numbers to remain stable. Use of JIT within the Toyota Production System means that individual cars can be built to order and that every component has to fit perfectly first time because there are no alternatives available. It is therefore impossible to hide pre-existing manufacturing issues; they have to be addressed immediately.

8. Kaizen (Continuous improvement)

A process that helps to ensure maximum quality, the elimination of waste, and improvements in efficiency.

Kaizen improvements in standardised work help maximise productivity at every worksite. Standardised work involves following procedures consistently and, therefore, employees can identify problem promptly.

Kaizen activities include measures for improving equipment, as well as improving work procedures, Literally 'good change', the word now refers to the culture and philosophy of continuously improving any department or functional process, thereby increasing productivity, quality and efficiency.

Within the Toyota Production System, *Kaizen* humanises the workplace, empowering individual members to identify areas for improvement and suggest practical solutions. The focused activity surrounding this solution is often referred to as a *kaizen blitz*, while it is the responsibility of each member to adopt the improved standardised procedure and eliminate waste from within the local environment.

9. Kanban (Pull System)

A tool used in the Toyota Production System to operate the

'pull' type production system.

It is a system that provides for the conveying of information between processes and automatically orders parts as they are used. Every item or box of items that flows through the production process carries its own kanban. Kanban are removed from items as they are used or transported and go back to the preceding processes as orders for additional items.

Though literally translated as 'signboard', the Toyota-developed method has become known as a clear, sign-based scheduling system triggering the logistical chain of production and maintaining it at an optimal level. Kanban is the quick-response system through which Just-In-Time production is achieved, harmonising inventory levels with actual consumption.

Toyota has six rules for the effective application of *Kanban*:

1) Never pass on defective products;

2) Take only what is needed;

3) Produce the exact quantity required;

4) Level the production;

5) Fine-tune production; and

6) Stabilise and rationalise the process.

Toyota Production System – Manufacturing Supermarket

A replenishment process which ensures that all manufacturing components ordered from outside suppliers are available to be loaded and delivered in one consignment. Derived from the system used by retail supermarkets, it levels the occasional spikes in demand experienced in individual factories by requiring suppliers to smoothly and systematically gather unusually large orders to a separate holding area, or 'virtual truck', ahead of the regular loading schedule. This process avoids any disruption to the tempo of deliveries and last-minute rushing around to complete an order.

10. Muda (Waste)

In management terms, refers to a wide range of non-value-adding activities. For example, anything an operator has to do within a process which does not add value but does add cost. Eliminating waste is one of the main principles of the Just-In-Time system. Waste incurs unnecessary finance costs and storage costs.

The first of three types of waste mentioned within the Toyota Production System (the others being *Mura* and *Muri*), the identification and reduction of which will reduce the unnecessary consumption of resources and increase profitability. Toyota divides Muda into seven resources that are frequently wasted:

1) Transportation — a cost that adds no value to the product but increases the risk of a product being damaged, lost or delayed;

2) Inventory — a capital outlay that if not processed immediately produces no income;

3) Motion — any damage inflicted through the production process, such as normal wear and tear in equipment,

repetitive stress injuries, or by unforeseen accidents;

4) Waiting — products that are not in transport or being processed;

5) Over-processing — when more work is done than necessary, or when tools are more complex, precise or expensive than necessary.

6) Over-production — larger batches or more products being made than is required.

7) Defects the loss involved in rectifying faulty parts or products.

11. Mura (Unevenness or irregularity)

Eliminating unevenness or irregularities in the production process is one of the main principles of the Just-In-Time system, the main pillar of the Toyota Production System.

The second of three types of waste mentioned within the Toyota Production System, notably identified and levelled through the application of Heijunka principles and Kanban devices.

Workflow is also smoothed by requiring members to operate multiple machines — also known as 'multi-process handling' — within any particular process, and by predicting and preparing for times of high demand.

12. Muri (Overburden)

Eliminating overburden of equipment and people is one of the main principles of the Just-In-Time system, the main pillar of the Toyota Production System. To avoid overburden, production is evenly distributed in assembly processes.

The third of three types of waste highlighted within the Toyota Production System, requiring the balancing of manufacturing pace to allow members sufficient time to achieve the correct standard of work. A reduced time frame will be too burdensome to achieve the objective, while the allowing of too much time is a waste of resource.

(Related: Takt Time [derived from German word Taktzeit, or 'cycle time'] — matching the pace of production with customer demand and the available work time).

13. Nemawashi

(Laying the groundwork or foundation)

The first step in the decision-making process. It is the sharing of information about decisions that will be made, in order to involve all employees in the process. During Nemawashi, the company seeks the opinion of employees about the decision.

Literally translated as 'going around the roots', particularly in the sense of digging around the roots of a tree to prepare it for transplant.

Within the Toyota Production System — and Japanese culture itself — the word has come to mean an informal process of laying the foundation and building a consensus of opinion before making formal changes to any particular process or project. Successful application of Nemawashi allows changes to be carried out with the consent of all parties.

14. Poka-Yoke (Mistake-proofing)

Failsafe devices in the production process (sensors, templates etc) that automatically stop the line when an abnormality occurs.

A Poka-Yoke is any part of a manufacturing process that helps a Toyota member avoid (yokeru) mistakes (poka). Its purpose is to eliminate defects by preventing, correcting, or highlighting errors as they occur – for example, a jig that holds parts for processing might be modified to only allow them to be held in the correct arrangement.

It is based on the premise of making work easier for workers. The objective is to thoroughly eliminate waste and shorten lead times to deliver vehicles to customers quickly, at a low cost, and with high quality. This production system is pursued in all areas of Toyota Motor Corporation, including vehicles and services, and all employees implement daily incremental kaizen.

THE TWO PILLARS OF TPS

The basic philosophy of the Toyota Production System is based on two pillars.

The first pillar is jidoka—

Which can be loosely translated as "automation with a human touch"—based on the concepts of stopping immediately when abnormalities are detected to prevent defective products from being produced and improving productivity to eliminate the need for people to be simply watching over machines.

The second pillar is Just-in-Time---

Based on the concept of synchronizing production processes—linking all plants and their production processes in a continuous flow—by making only what is needed, when it is needed, and in the amount needed.

These two pillars enable the production of vehicles that satisfy customer requirements quickly, at a low cost, and with high quality.

Jidoka

Full application of Jidoka means that the process which created any issue is subsequently evaluated to remove the possibility of re-occurrence.

Jidoka (English: Autonomation — automation with human intelligence): The principle of designing equipment to stop automatically and to detect and call attention to problems immediately whenever they occur (mechanical *jidoka*). In the Toyota Production System, operators are equipped with the means to stop the production flow whenever they note anything suspicious (human *jidoka*), thereby preventing the waste that would result from producing a series of defective items. It also liberates operators from controlling machines, leaving them free to concentrate on tasks that enable them to exercise skill and judgement instead of monitoring each machine continuously.

Employing *Jidoka* principles throughout the production process is a vital element of the Toyota Production System, forcing imperfections to be immediately addressed by self-inspecting workers and thereby reducing the amount of work added to a defective product. Some automated machines can also function in the detection process, allowing human operatives to only be engaged when alerted to a problem.

Jidoka in the TPS is "automation with a human touch," where human wisdom is added to automation. Human wisdom means that when an abnormality occurs, such as a machine or equipment abnormality, quality abnormality, or a work delay, the machine or equipment can detect the abnormality and stop automatically, or the operator can stop the line by pulling the stop cord themselves. This eliminates the outflow of defective products while also making it possible to build quality into processes by clearly detecting abnormalities and preventing them from recurring. Furthermore, having the ability to stop when an abnormality is detected means that machines and equipment no longer need to be watched over, saving labour by reducing working hours.

To create these kinds of machines, it is necessary to first be able to do the work smoothly and correctly by hand, determine abnormalities in the work, and replace those operations with machines.

In other words, rather than starting with a machine from the beginning, you must first try doing it thoroughly by hand, implement kaizen, eliminate waste, inconsistencies, and unreasonable requirements—known respectively in Japanese as *muda, mura, muri*—and make it possible for anyone to do the work. You must then make it possible to detect abnormalities in the work and build that into the actual machines. These incremental efforts lead to a production line that is high-quality, low-cost, flexible, and easy to maintain. This implementation of kaizen on work is the bedrock of jidoka. It doesn't matter how much machines, robots, or IT excel; they can't evolve any further on their own. Only humans can implement kaizen for the sake of evolution. In other words, craftsmanship is achieved by discovering the basic principles of manufacturing through manual work and then applying them on the production line to steadily implement kaizen. This cycle of kaizen in both human skills and technologies is critical for taking on the challenge of new technologies and construction methods. Human wisdom and ingenuity are indispensable to delivering ever-better cars to customers. Going forward, we will maintain our steadfast dedication to constantly developing human resources who can think independently and implement kaizen.

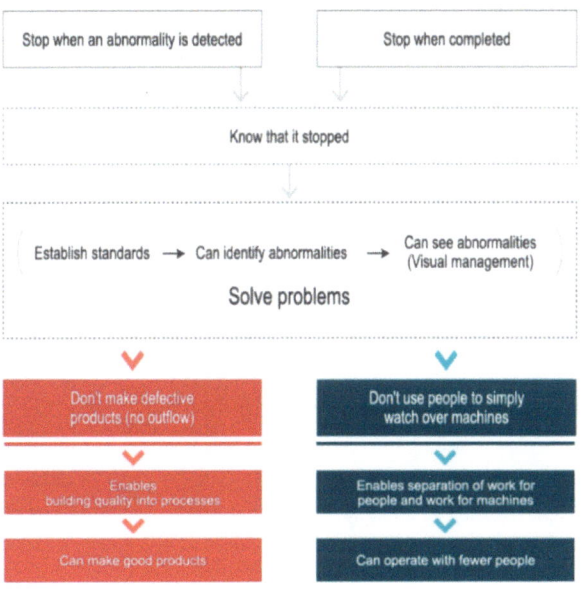

JUST-IN-TIME

- Making only what is needed, when it is needed, and in the amount needed
- Fulfilling orders from customers as quickly as possible

A car is made up of more than 30,000 parts. These parts are made not only by Toyota but also at the plants of many of our business partners. All plants must work with complete synchronization to make the vehicles quickly and without waste. All adhere to the following principles of Just-in-Time to achieve synchronized production:

1) Only make what is needed by the customer, when it is

needed, and in the amount needed;

2) Don't allow goods and information to be held up during production;

Just-in-Time (JIT) JIT is developed and perfected by Taiichi Ohno and practiced at Toyota to meet customer demands. This management philosophy aims at providing the customer with exactly what is required, at the right time, right place and at right cost. The goal of JIT is to minimize the presence of NVA activities in the processes. Eliminating unwanted steps will naturally result in shorter cycle time, better on-time delivery (OTD) performance, optimum resource utilization, lower defects, lesser cost and greater profits! JIT is also known as inventory-less production as the output is just what the customer demand is. JIT requires Kanban to be operational in the system that signals information from upstream processes on customer requirements. JIT also requires a balanced load so that there are no bottle necks in the process.

Just-in-Time (JIT) Some ways to implement

JIT in Service Industry are Standardized Work – Standard Operational Instructions Kanban – Signal from upstream process steps at the right time Stabilized Schedules Uniform Work Load Focused Work Groups Periodic review of Value Stream and elimination of NVAs Kaizen – Continuous measurement and improvement in processes Key to Value Delivery –

"Right Need, Right Time, Right Price"

Make them at the pace at which they're sold. It would take many months to fulfil a customer's order if all the parts were made only after receiving it. To avoid that, the minimum number of parts needed are stocked in advance on the vehicle assembly line so that a car can be built as soon as the order is received. The preceding process has a store of finished products from which the next process can pick up the parts that it needs. The preceding process is also stocked in advance with the minimum number of parts needed to re-make parts picked up by the next process before the next pick-up, allowing it to immediately replenish whatever was picked up. Having all processes engaged in this loop achieves wasteless production where we only make what is needed, when it is needed, and in the amount needed to fulfil customer orders and ensure that only sellable items are produced.

The continuous pursuit of these two pillars of the TPS are the wellsprings of Toyota's competitive strength and unique advantages. We will develop human resources throughout Toyota who put this philosophy into practice to make ever-better cars that will be cherished by customers.

Just-in-time (JIT) production is a 'pull' system of providing the different processes in the assembly sequence with only the kinds and quantities of items that they need and only when it needs them. Production and transport take place simultaneously throughout the production sequence – inside and between all the processes.

The primary objectives of just-in-time production are to save warehouse space and unnecessary cost-carrying and to improve efficiency, which means organising the delivery of component parts to individual work stations just before they are physically required.

To apply this flow efficiently means relying on ordering signals from *Kanban* boards or by forecasting parts usage ahead of time, though this latter method requires production numbers to remain stable.

Use of JIT within the Toyota Production System means that individual cars can be built to order and that every component has to fit perfectly first time because there are no alternatives available. It is therefore impossible to hide pre-existing manufacturing issues; they have to be addressed immediately.

5 ,S' HOUSEKEEPING

The 5 S' Approach

"5 S" is one of the most important concept in Lean. Originally formulated in Japan for streamlining the housekeeping concept, it can be applied to physical work environment, database, table spaces, shop floors and any other area that has a tendency to accumulate waste and result in reduced efficiency. Thus, "5 S" focuses on effective workplace organization and standardized work procedures. It simplifies work environment, reduces wastes and improves quality, efficiency and safety.

The 5S' is a systematic Approach to housekeeping and organisation in the workplace.

It aims to

- Remove waste from the workplace.
- Provide reduction in non-value-added activities.
- Provide an environment where continuous Improvement is embraced
- Improve safety
- Increase quality.

Thus, "5 S" focuses on effective workplace organization and standardized work procedures. It simplifies work environment, reduces wastes and improves quality, efficiency and safety.

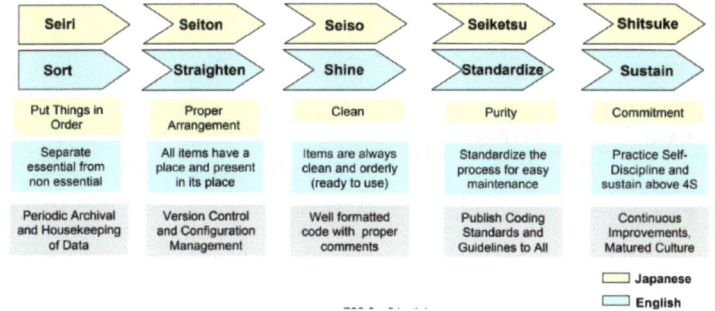

The 'S' are five Japanese words, which provide the steps of improvement

Step I (Seirl)
Clear out and Classify Identify

- What you KNOW you definitely need all the time.
- What you KNOW you definitely need occasionally. What you THINK you might need in the future.
- What you THINK you no longer need.

Never Assume. Ensure that everyone is in agreement. Check to make sure that what you Think becomes what you Know Remember!

"Redundant Items cost money to store-If you don't need them, don't keep them"

Step 2 (Sefton)
Configure

Put everything in its correct place.

- Organize storage for all equipment i.e. what, where, how many.
- Frequency of use defines location relative to point of use.
- If it's used frequently keep it close /low frequency store offline.
- Shop organization must have clear walkways, work areas, rest areas, etc.
- Consider using different coloured floor areas to highlight walkways work storage area etc.
- All essentials Such as fire extinguisher and safety equipment must be visible and easily accessible.

The place must be:

- Appropriate to usage
- Well maintained
- Tools Easily located
- Tools I locations Clearly identified

"A place for everything and everything in its place! "

Step 3 (SoIso)

Clean and Check

Regular cleaning schedules are required.

- Create a clean and tidy working environment and maintain it.
- Define responsibility zones for cleaning areas. and

clarify roles and responsibilities.

- Develop regular routines for maintaining a clean environment (e.g. 5min 5 S etc).

Cleaning must become an activity which is:

- Ongoing
- Monitored
- A source of pride
- Seen as a value-added contribution

Cleanliness is the basis of quality. Once the workplace is clean it must be maintained.

Creating a spotless workplace

Five Principles of Lean

What more can be improved?

1. Establish Pull → What's the bare minimum produce needed

2. Continuous Improvement → How can processes get "Leaner"

3. Map the Value Stream → Information/material that the customer is willing to pay for

4. Identify Value Stream → What are the chain of steps that create value

5. Create Flow → What are the wastes to be removed to create more value.

Step 4 (Seiketsu)
Conformity
Good Housekeeping.

Decide what is classed as an abnormality and make it visible to the operator. Design clear, standard labels for locations, tools, machine conditions and locate them in standard positions.

Provide indicators where limits can be exceeded. Draw position markers in places where items are removed and returned.

Standards of cleanliness.

Procedures for maintaining standards.

- Standard marking and labelling of all items. Standard methods for indicating limits, identifying locations, etc.
- The system must be controlled and maintained.

Step 5 (Shitsuke)
Custom and Practice
Adhere to the system rules

- Develop and maintain habits through training and discipline. (At all time)
- Use Visuals rather than verbal communication to train

for new procedures.
- Involve everyone in the development of standard documentation. (e.g. Check sheets etc.)
- Be conscious of time (3-10min 5 S).

It is important to maintain discipline.
- Incorrect practices must be identified.
- Correct practices have to be taught and demonstrated.
- Ensure everyone's understanding and agreement.

"Without training and discipline the other steps will fail"

THE 5 'S'

We all want to work in the best possible environment.

- Good housekeeping is important as it:
- Creates an environment in which even minor abnormalities and mistakes will be obvious
- Produces an easily managed, safer and more pleasant environment
- Portrays professionalism and efficiency to others (particularly customers)
- It can be expected that the standards displayed in the environment will be reflected in the product
- Stimulates efforts to improve productivity through better use of people, space, equipment, time and materials.

5. Lean Waste

LEAN WASTE

Anything which adds cost without value of time and money or Product and Services called Waste

FORM OF WASTE

1. Bad quality defective parts (Rework / Rejection)
2. Unnecessary Movements
3. Over Production
4. Unnecessary transportation
5. Extra Inventories
6. Over Processing
7. Waiting

WASTES OF LEAN MANUFACTURING

1) Transportation – In production, it may mean moving parts and materials from one place to another
2) Inventory – Undelivered products or parts. Overstocking with equipment that may be in need somewhere in the future
3) Motion – Unnecessary movement of employees or machinery
4) Waiting – Waiting for goods to be delivered
5) Overproduction – Too many items produced "just in case"
6) Over-processing – Spending a lot of time on a given task. Adding a feature that doesn't bring value
7) Defects – Broken parts or defective parts that need to be reworked

WASTES IN SOFTWARE DEVELOPMENT

1) Transportation – Switching between tasks too often, countless interruptions from colleagues.
2) Inventory – Undelivered code or undelivered features
3) Motion – Unnecessary meetings or extra effort to find information
4) Waiting – Waiting for testing to complete, waiting for code review, and so on
5) Overproduction – Producing features that nobody is going to use
6) Over-processing – Unnecessary complex algorithms solving simple problems
7) Defects – Bugs

Wastes in Marketing

1) Transportation – Task switching, interruptions, unnecessary long marketing funnel
2) Inventory – Fully-prepared marketing campaigns which stay unlaunched. Licensed tools that nobody uses
3) Motion – Unnecessary meetings, extra effort to find information, attending events without clear agenda
4) Waiting – Waiting for approval from higher management
5) Overproduction – Performing many different marketing activities without having a clear vision and strategy
6) Over-processing – Generating countless marketing reports manually, while they can be automated
7) Defects – Wrong brand communication, mistakenly

Wastes in Project Management

1. Transportation – Task switching, interruptions, unnecessary long marketing funnel
2. Inventory – Purchased online tools that teams rarely use, office supplies that exceed needs.
3. Motion – Badly structured workspaces and lack of organizational paths, too many meetings, extra effort to find information, and so on
4. Waiting – Waiting for approval from higher management
5. Overproduction -Filling unnecessary great amount of documents
6. Over-processing – Multiple levels of approval for small tasks
7. Defects – An incorrect collection of data

THE 7 WASTE

1. ## Waste - Motion / Transportation

- Long travelling distances consume time. This results in long lead times, reduces response to customer demands, transportation damage, lost components and makes it much harder to manage.
- Effort to close couple resources must be applied

2. ## Waste - Motion / Transportation

- Long travelling distances consume time. This results in long lead times, reduces response to customer demands, transportation damage, lost components and makes it much harder to manage.

- Effort to close couple resources must be applied ----.the same principle applies to non- manufacturing areas

3. Waste - Waiting Time

- Waiting or queuing is waste since it does not add value to the product
- Waiting manifests itself in the form of inventory accumulations at process stages. High inventory in turn encourages high product waiting times.

4. Operator waiting time implies under-utilisation and poor control of workflow(Waiting results in)
- Long lead times
- Wasted floor space
- Increased damage
- Potential obsolescence
- Misplaced items
- Demoralized workforce
- Poor workflow continuity
- Ineffective use of time
- Reduced competitiveness
- Ineffective Production
- Planning / Control
- Where's my next job ??

5. Waste – Overproduction

- The production of goods in excess of absolute Customer requirements

- Manufacturing too much, too early or "Just in Case".
- Overproduction discourages a smooth flow of goods or services.
- Takes the focus off what the Customer really wants.
- Leads to excessive Inventory.

6. Waste - Processing Time

- Ensuring processing/ manufacturing time is optimized.
- Target
- To add value to the product with the least effort.
- By improving processing efficiency, we ultimately commit less resources to achieve the same customer satisfaction
- Reduced lead-time (reduced cycle time)
- Improved customer response Flexibility
- Outcome
- Greater output / productivity
- Reduced defects
- Improved resource utilization

7. Waste – Defects

- Defects reduce or discourage customer satisfaction
- Defects have to be rectified
- Rectification costs money with regard to time, effort and materials.
- Defects in the field will lose customers.
- Right First Time is the Key.

Tools Use to Tackle the 7 Forms of Waste

- There are different tools in Lean you can use to identify and

eliminate wasteful activities. They will guide you through the work process you want to examine and show you the weak spots.
- Probably, the most appropriate tool for identifying Muda is the Gemba walk. It is a technique that allows you to go and see where the real work happens. This way, you can observe different processes in action and see where wasteful activities appear.

6.Tools And Technique- Kanban

KANBAN

Kanban is the basis of
"PULL SCHEDULING"
which seeks to ensure that
preceding operations only supply
or make as many items as are required by the following operations.

Kanban

Kanban is one of the key tools of Just in Time (JIT) and Pull systems. It is a Visual Communication Media that acts as a signal to the operator or between processes that communicates the demand at the right time to the right process step. Kanban helps organizations maintain an efficient flow of output throughout the entire process value stream with low inventory and work in process. It also caters to seasonal fluctuations in demand. In manufacturing environment, the demand is communicated to the production units by a printed card that contains information on part name, description and quantity. Nothing is manufactured unless there is a "signal" to manufacture. This is in contrast to a push manufacturing environment where production is continuous. In Service and IT environment, electronic communication and agreement are essential before initiating any work (eg: Purchase Order, Requirement Sign-off). There are two types of Kanban – Production and Withdrawal for initiating or stopping the work.

Just-in-time (JIT) production is a 'pull' system of providing the different processes in the assembly sequence with only the kinds and quantities of items that they need and only when it needs them. Production and transport take place simultaneously throughout the production sequence – inside and between all the processes.

The primary objectives of just-in-time production are to save warehouse space and unnecessary cost-carrying and to improve efficiency, which means organising the delivery of component parts to individual work stations just before they are physically required.

To apply this flow efficiently means relying on ordering signals from *Kanban* boards or by forecasting parts usage ahead of time, though this latter method requires production numbers to remain stable.

Use of JIT within the Toyota Production System means that individual cars can be built to order and that every component has to fit perfectly first time because there are no alternatives available. It is therefore impossible to hide pre-existing manufacturing issues; they have to be addressed immediately.

Differences in Push and Pull Systems (Production)

Visual representation of a Pull Vs Push System

Pull
- ✓ Smooth Flow
- ✓ Reduced Cost
- ✓ No Inventory
- ✓ Just in Time

Push
- ✓ Forced Flow
- ✓ High Operational Cost
- ✓ High Inventory
- ✓ Rejected Output

Be sensitive to Customer needs, Cater to exact requirement!

Pushing System
1. Pushing Production
2. (traditional Western production system)
3. This process pushes work along the system:
4. Flow is controlled by the production plan
5. Work is pushed on to work centres, regardless of the level of work in progress (WIP) at each of the work centres.
6. Batch sizes may not be related to demand
7. Assembly Products Warehouse
8. All the plans are based on forecasts

Pull Systems Production
1. Pulling Production

2. This process pulls work along the system in line with demand.
3. Algorithm Flow is controlled by customer requirements.
4. Kanban authorizes upstream processes to re-manufacture parts when they are withdrawn into downstream processes.
5. Small batch sizes and smoothing of demand limits WIP.

Parts →Sub- Assemblies→ Assemblies →Customer

Production Plan→Purchasing Plan→Processing Plan

Assembly Plan→Delivered to→Supplier→Material Warehouse→Stock in Process→Sub Assembly→Assembly →Products Warehouse stock

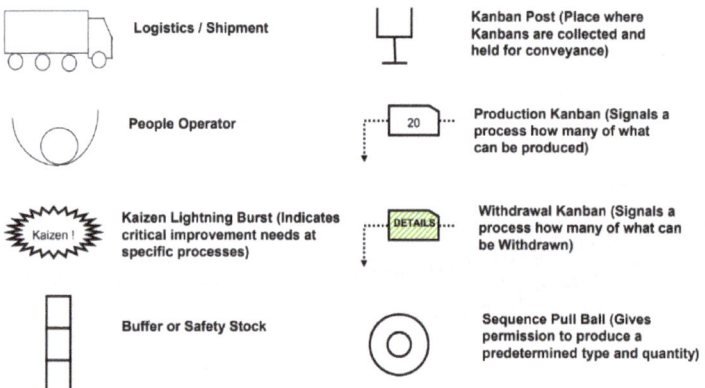

The Function of Kanban

1. Production control (indicator of when, how many, and what to produce)
2. Material control (controls the movement of parts and flow of information)
3. Improvement tool (reduces inventory, reduces lead- time and encourages visual control)
4. Focuses on JIT, Reducing lead time
5. Kanban reduces the level of paperwork in the system

6. Differences in Push and Pull Systems Marketing
7. There are several variations of the Kanban System being used at the present time.
8. How ever one thing they all have in common is that they can be adapted and developed for most types of production systems.

6. Typical variants of Kanban

   Footprint→Max→Minimum / maximum→Stock levels→Min →2 Bin→Card System→Kanban Card

   ### Kanban Benefits

   - ✓ Synchronization of supply and demand
   - ✓ Totally customer driven demand
   - ✓ Optimum inventory
   - ✓ Defect detection time reduced
   - ✓ Better machine utilization
   - ✓ Reduced or eliminated queues
   - ✓ Problem identification leading to resolution
   - ✓ Lead time is kept to a minimum
   - ✓ Simple and cheap to operate

   Supplier→Kanban Re-Order→Stores→Customer

   ## Kanban Re-Order

   Optimizing software development with Kanban flow Kanban flow, a cornerstone of agile and DevOps Methodologies, drives efficiency by orchestrating seamless task progression through visualized workflows. Kanban flow mirrors the streamlined inventory management of supermarkets, ensuring tasks move through development processes precisely when needed. Visualized on Kanban boards, tasks represented as cards enable transparent progress tracking and swift identification of bottlenecks. By limiting work-in-progress (WIP), teams optimize resource allocation and maintain a steady workflow. Kanban's focus on continuous improvement is facilitated by metrics like. control charts and cumulative flow diagrams, empowering teams to refine workflows iteratively.

In software development kanban flow fosters dynamic task management, accelerates delivery cycles, and enhances customer satisfaction through focused, uninterrupted work. In essence, kanban flow epitomizes efficiency—a harmonious blend of transparency, adaptability, and continuous improvement—unlocking the full potential of agile methodologies.

Structuring your Kanban flow

Establishing a structured kanban flow within your software development team is essential to implementing kanban differently. This ensures smooth task progression and optimized workflow management.

How you construct your kanban flow

Visualize workflow: Begin by visualizing your team's workflow on a Kanban board. Whether physical or virtual, the board should depict each stage of the development process, from task inception to completion.

Standardize workflow: Define and standardize the workflow stages according to your team's processes and requirements. Common stages include "To Do," "In Progress," and "Done," but customize as needed to reflect your unique workflow.

Identify blockers and dependencies: Ensure that your kanban board enables immediate identification of blockers and dependencies. This transparency allows for prompt resolution and prevents workflow disruptions.

Set work-in-progress (WIP) limits: Implement WIP limits for each workflow stage to avoid overburdening and to maintain a steady workflow. WIP limits help optimize resource allocation and reduce multitasking, fostering higher productivity.

Encourage collaboration: Foster a culture of collaboration within your team, where members collectively address bottlenecks and work together to ensure smooth workflow progression. This collaborative approach promotes efficiency and accelerates task completion.

Utilize kanban cards: Represent each task as a kanban card on the board, containing essential details such as task description, assignee, and estimated time for completion. Kanban cards facilitate visual tracking of task progress and promote transparency within the team.

By structuring your kanban flow in this manner, you can streamline your software development processes, enhance team collaboration, and maximize efficiency in task management.

Exploring the origins of Kanban

Kanban is prominent among today's agile and DevOps software teams, but the Kanban methodology dates back more than 50 years. In the late 1940s, Toyota began optimizing its engineering processes based on the same model supermarkets used to stock their shelves.

Kanban Card

Supermarkets stock just enough inventory to meet consumer demand, a practice that optimizes the flow between the supermarket and the consumer. Because inventory levels match consumption patterns, the supermarket gains significant efficiency in inventory management by decreasing the excess inventory it must hold at any given time. Meanwhile, the supermarket can still ensure that essential products are always in stock.

When Toyota applied this same system to its factory floors, the goal was to better align its massive inventory levels with the actual consumption of materials. To communicate capacity levels in real-time on the factory floor (and to suppliers), workers would pass a card, or "kanban," between teams.

When someone emptied a bin of materials used on the production line, a kanban card was passed to the warehouse describing what material was needed, the exact amount of this material, and so on. The warehouse would have a new bin of this material waiting, which they would then send to the factory floor and, in turn, send their own kanban to the supplier. While this process has evolved since the 1940s, this same "just in time" (JIT) manufacturing process remains at the heart of the kanban methodology.

Kanban for software teams

Agile software development teams today can leverage JIT principles by matching the amount of work in progress (WIP) to the team's capacity. This gives teams more flexible planning options, faster output, clearer focus on continuous improvement, and transparency throughout the development cycle.

Kanban is one of the most popular software development methodologies agile teams use today. Kanban offers additional advantages to task planning and throughput for teams of all sizes.

Planning Flexibility

A kanban team focuses only on the work that's actively in progress. Once the team completes a task, they select the next task from the backlog. The Product Owner is free to reprioritize work in the backlog without disrupting the team because any changes outside the current work items don't impact the team.

As long as the product owner keeps the most important work items on top of the backlog, the development team can rest assured they are delivering maximum value back to the business.

Savvy product owners always engage the development team when considering changes to the backlog.

For example, if user stories 1-6 are in the backlog, user story

6's estimate may be based on the completion of user stories 1-5. It's always a good practice to confirm changes with the engineering team to ensure no surprises.

Shortened Time Cycles

Cycle time is a key metric for Kanban teams. Cycle time is the amount of time it takes for a unit of work to travel through the team's workflow—from the moment work starts to the moment it ships. By optimizing cycle time, the team can confidently forecast the delivery of future work. Over lapping skill sets lead to shorter cycle times

When only one person holds a skill set, that person becomes a bottleneck in the workflow. Therefore, teams employ best practices like code review and mentoring to help spread knowledge. Shared skills mean team members can take on heterogeneous work, further optimizing cycle time.

Additionally, this approach empowers the entire team to address any work bottlenecks collectively, facilitating a swift resolution and ensuring a smooth workflow. For example, testing responsibilities extend beyond QA engineers to include developers, fostering a collaborative effort to maintain efficiency. In a kanban system, the entire team ensures work moves smoothly through the process.

Fewer Bottlenecks

Multitasking kills efficiency. Increased workload simultaneously leads to more frequent impeding the progress of tasks toward completion. That's why a vital tenet of the kanban process is limiting the work in progress (WIP). Work-in-progress limits highlight bottlenecks in the team's process due to a lack of focus, people, or skill sets.

For example: a typical software team might have four Workflow states.

To Do, In Progress, Code Review, and Done. They could set a WIP limit of 2 for the code review state. That might seem like a low limit, but there's a good reason for it. Developers often prefer to write new code rather than spend time reviewing someone else's work. A low limit encourages the team to pay special attention to issues in the review state and review others' work before raising their code reviews, ultimately reducing the overall cycle time.

Visual metrics

One of the core values is a strong focus on continually improving team efficiency and effectiveness with every iteration of work. Charts provide a visual mechanism for teams to ensure they're continuing to improve.

When the team can see data, it's easier to spot bottlenecks in the process (and remove them). Two common reports Kanban teams use are control charts and cumulative flow diagrams. A control chart shows the cycle time for each issue and a rolling average for the team.

The team's goal is to reduce the time an issue takes to move through the entire process. Seeing the average cycle time drop in the control chart indicates success.

A cumulative flow diagram shows the number of issues in each state. The team can easily spot blockers when the number of issues increases in any given state. Issues in intermediate states such as "In Progress" or "In Review" are not yet shipped to customers, and a blocker in these states can increase the likelihood of massive integration conflicts when the work gets merged upstream.

Continuous delivery

Computer Delivery (CD) describes the process of releasing work to customers frequently. Continuous integration is the practice of automatically building and testing code incrementally throughout the day. Together, they form a CI/CD pipeline that is essential for DevOps teams to ship software faster while ensuring high quality.

Kanban and CD beautifully complement each other because both techniques focus on the just-in-time (and one-at-a-time) delivery of value. The faster a team can deliver innovation to the market, the more competitive their product will be.

Kanban teams focus on precisely that: optimizing the flow of work out to customers.

Scrum vs. Kanban

Kanban and Scrum share some of the same concepts but have very different approaches.

Agile

Agile is an iterative approach to project management and

software development that helps teams deliver value to their customers faster and with fewer headaches. Agile methodologies are immensely popular in the software industry since they empower teams to be inherently flexible, well-organized, and capable of responding to change.

7. Tools and Technique -Kaizen

KAIZEN

Kaizen is the Japanese key to success.

Kaizen is split as Kai and Zen with Kai = Change; Zen = Good. Thus, Kaizen translates to "Change for the Better"! This is also the first step for Continuous Improvement. Kaizen also means "To take it apart and put back together in a better way". A process, product, system or service is taken apart to put back in a better way The goal of Kaizen is to remove waste, standardize work and deliver right product at right time with optimum resources to customers. This is achieved by participation by all stakeholders, people from all levels in an organization as well as third parties, if part of a value stream. Kaizen moves ahead with one small step at a time, validate the improvement and take the next step. The Kaizen Cycle is similar to Deming Cycle of Plan-Do-Check-Act. Typical Kaizen cycle includes the following Define a Standardized Value Stream Measure and Analyze each step in the Value Compare against requirements Stream in terms of process and wait Time Innovate and make incremental improvements to meet requirements and Increase productivity Standardize the new, improved Value Stream Continue analysis and improvement on the Value Stream of the Process / Product / Operation or Service.

Kaizen is a Japanese word meaning "change for better." And the original meaning of the Japanese word "Kaizen" from the Shogakukan Dictionary could be literally translated as "The act of making bad points better."

While that is true, there is much more to it. Regardless of what you might have heard before, continuous improvement is not the sole definition of Kaizen. Instead, it is the result of it. In fact, the literal translation of continuous improvement in Japanese is "Kairyo".

Kaizen is more of an internal process that happens within your own mind. The goal is to realize your potential, break the status quo, and, this way, achieve improvement. With that being said, a more precise way to define Kaizen would be "continuous self-development."

Actually, the modern sense of the word originated in Toyota factories. After WW2, many Japanese businesses were influenced by the methodologies brought by American advisors sent as part of the Marshall Plan.

Although this practice was implemented elsewhere, Toyota is the brightest example of a company that made an excellent practice of continuous improvement, creating effective management systems to generate, capture, and review improvements in never-ending cycles.

Tools and Techniques

1) Kaizen Rules Work on process and results (not results only) Systemic thinking (big picture, not solely the narrow view) Non-judgmental and Non-blaming (blaming is wasteful) All the stakeholders irrespective of their level should participate and contribute

When and How does Kaizen get initiated?

The team has to continuously analyze and evaluate the activities of their operational process. Any activity that has a scope of improvement to reduce cycle time will trigger a Kaizen Event. The above-mentioned Kaizen Cycle and Rules are to be adhered to during the event to get maximum benefit out of it. The next level of Improvement Activity is "Kankaku", meaning "Break-through Change". These are relatively rare and focus on radical change in systems and structures.

5 Why- In Problem Solving

If we have the data for analysis, the root cause may be obvious. But when we do not have data in hand, we use certain tools and techniques to understand the problem and the cause.

"5 Why" is one of the tools used to drill down to the root cause of a problem.

"5 Why" is often used in conjunction with Cause and Effect Diagram and generates thoughts on the problem and causes by applying Brainstorming technique.

Basic Guidelines for 5 Why Exercise

Define the problem clearly to all stakeholders Brainstorm the cause of problem - "Why does the problem exists" The answer to the first "why" will prompt another "why" and so on. Continue asking "Why" 5 times (or more / or less) till the "Root Cause" is arrived and agreed by all stakeholders.

Look for the next tool to eliminate the root cause and the problem

Thus, "5 Why" strategy is simple and easy to learn and practice. It is often very effective in uncovering the causes of long term problems. It can be adapted quickly and applied almost to any problem.

5 Why Example

Problem: Customer is unhappy and escalated the issue on late deliveries ?

1.Why does the problem exists?

! Its due to internal external environments.

2.Why were the deliveries late?

! Because the job took longer than expected ?

3.Why did the job take longer than planned?

! Because the complexity of the job was not estimated correctly ?
4.Why was the complexity not estimated correctly?

! Because the scope and complexity of the work was not fully analyzed at the time of estimation ?

5.Why was the work not analyzed completely before finalizing estimation?

! Because we were running behind schedule of other projects and also did not want to lose this project Here, it is evident that we need to review the 'estimation guidelines' or 'shortened schedule due to customer pressure'

'Attack the Cause, Not the Symptom"

Toyota's overall system of techniques for production management is called the Toyota Production System (TPS). The system rests upon several core principles, one of which is labelled Kaizen.

For Toyota's usage (or generally, any manufacturing usage), it largely means continuous improvement through the act of self-development. Kaizen became one of the core practices behind

Lean manufacturing in the USA and later in Lean management.

Nowadays, in the modern dynamic and unsafe business environment, more and more companies are on the way to becoming more efficient by applying Lean Thinking. A crucial part of it is the Kaizen culture.

Kaizen Philosophy and Culture in Practice.

To achieve Kaizen, you need to adopt the practice of self-criticism. In Japanese, that practice is known as "Hansei." This means that you need to hold yourself accountable and find room for improvement, even if everything is going according to plan.

Adopting this type of mindset will give you the ability to break the status quo and push yourself to the limits. While positive thinking will show you everything as a success, it is the negative emotion of "it could've been better" that will give you the motivation to improve and eventually conquer new peaks continuously.

For example, in the Toyota production system, all line personnel is expected to stop the moving production line in case an abnormality occurs. Then, the staff, alongside their supervisors, suggest an option to resolve the abnormality.

Furthermore, when the project ends, a Hansei-kai (reflection meeting) is held to analyze the entire process, including any abnormalities that have occurred. Here, it is important to mention that a Hansei-kai process would happen even if the project were finished successfully, with no issues found along its lifecycle.

These approaches initiate Kaizen and usually deliver small improvements during the process and at the end of it. As a result, the culture of continually aligned small enhancements and standardization leads to significant overall productivity improvement changes.

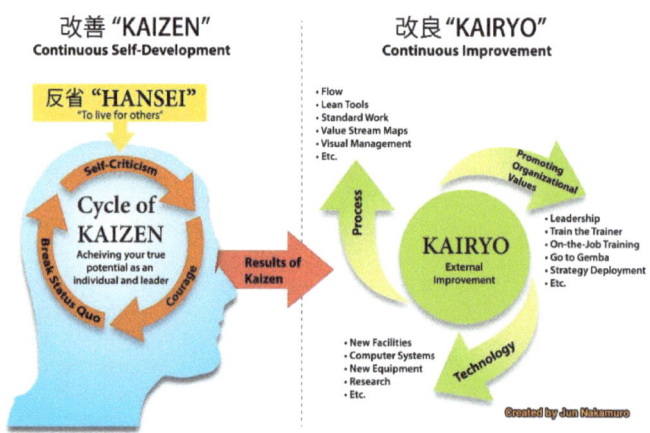

Image credit: Image credit: Nakamuro Jun

The developed Kaizen methodology includes making changes and monitoring results, then adjusting. Large-scale planning and extensive project scheduling are replaced by smaller experiments, which can be quickly adjusted when new improvements are suggested.

Kaizen Work

The Kaizen philosophy is rooted in the idea that improvement is a constant pursuit and there is always room for growth. Moreover, it upholds the principle of Respect for People. The Kaizen approach involves identifying areas for improvement, finding effective solutions, and implementing them. Furthermore, the process is iterative, and any issues that are not adequately resolved are revisited in subsequent cycles. A seven-step cycle can be employed to facilitate continuous improvement, providing a structured methodology for executing this process.

The 7-Step Kaizen Process (Cycle for Continuous Improvement)

The implementation of Kaizen can be accomplished through a seven-step cycle that aims to establish a culture of continuous improvement. The process involves the following steps:

1. Foster employee involvement. Encourage the participation of employees by empowering them to identify issues and problems. This approach generates a sense of ownership in the change process. Often, this is achieved by creating specific teams responsible for collecting and sharing information throughout the organization.

2. Identify problems. Utilize feedback from all employees to identify areas that require improvement or present potential opportunities. If there are multiple issues, create a comprehensive list to address each one systematically.

3. Develop a strategic solution. Foster a culture of innovation by encouraging employees to generate creative ideas. Evaluate all ideas, and select one or more winning solutions to implement.

4. Pilot the solution. Involve all stakeholders in the implementation of the chosen solution, and initiate pilot programs or other small-scale tests to assess its effectiveness.

5. Evaluate the outcomes. Regularly monitor progress and engage ground-level workers to ensure their continued involvement. Evaluate the success of the implemented change and identify opportunities for improvement.

6. In the event of positive outcomes, it is recommended to implement the solution organization-wide.

7. It is advisable to continuously repeat these seven steps to test new solutions as needed or address new problem lists.

Kaizen Event: Meaning and Example

A Kaizen Event, also known as a "Kaizen blitz", is a purposeful, time-limited initiative designed to facilitate an efficient brainstorming session focused on a single problem and enhancing an existing process.

Aligned with the principle of and Six Sigma, a Kaizen event is cantered on the concepts of focus and speed. Organizations can achieve significant breakthroughs that lead to swift process improvements by assembling an appropriate team to address a particular issue.

Typically, a Kaizen Event spans three to five days. Despite its brevity, this approach should be incorporated into a business's long-term process improvement plans as a regular practice.

A common process for a Kaizen event typically involves the

following steps:
1. Establish clear objectives and provide relevant context.
2. Evaluate the current state and devise a plan for making enhancements.
3. Implement the identified improvements.
4. Analyse the results and address any issues that arise.
5. Present the outcomes and identify any required follow-up actions.

This type of Kaizen process cycle is commonly known as PDCA (Plan, Do, Check, and Act) and is widely used to improve business operations. This approach brings a scientific and methodical approach to making enhancements and achieving desired outcomes.

PLAN: Plan your improvements, including setting goals.

DO: Put in place the actions required for improvement.

CHECK: Measure your success relative to your baseline.

ACT: Adjust or tweak your changes.

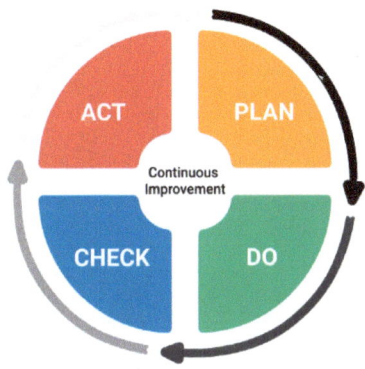

As you progress through each step, you keep the wheel

moving, representing continuous improvement.

When you arrive again at your baseline stage, you consider your previous developments and plan the next ones.

But remember, this is just the result of Kaizen. Image credit: Jun Nakamura

The developed Kaizen methodology includes making changes and monitoring results, then adjusting. Large-scale planning and extensive project scheduling are replaced by smaller experiments, which can be quickly adjusted when new improvements are suggested.

How Does Kaizen Work?

The Kaizen philosophy is rooted in the idea that improvement is a constant pursuit and there is always room for growth. Moreover, it upholds the principle of Respect for People. The Kaizen approach involves identifying areas for improvement, finding effective solutions, and implementing them. Furthermore, the process is iterative, and any issues that are not adequately resolved are revisited in subsequent cycles. A seven-step cycle can be employed to facilitate continuous improvement, providing a structured methodology for executing this process.

The 10 Principles of Kaizen

The successful execution of the Kaizen philosophy requires fostering the appropriate mindset throughout an organization. To address this, ten principles that embody the Kaizen mindset have emerged as central to the philosophy. These principles include:

1. A band owning preconceptions.
2. Taking initiative in problem-solving.
3. Rejecting the status quo.
4. Embracing a mindset of iterative, adaptable change instead of striving for perfection.
5. Viewing mistakes as opportunities for solutions.
6. Creating an inclusive environment where all employees feel empowered to contribute.
7. Challenging apparent issues by asking "why" five times to ensure a proper root cause analysis.
8. Gathering information and perspectives from multiple sources.
9. Using creativity to identify low-cost, incremental improvements.
10. Never stop continuously improving.

What Are the Advantages and Disadvantages of Kaizen?

There are numerous advantages of implementing Kaizen in an organization. Nevertheless, it is important to acknowledge that there are certain scenarios where it may not be the most fitting approach. Below are some of the advantages and disadvantages of implementing Kaizen.

The Advantages of Kaizen

1. Kaizen helps everyone to speak the same language: Small continual changes and standardization almost automatically take everyone on the same page. Employees are part of the process and its creation, improving themselves along with it.

2. Kaizen creates a growth mindset: Company values come between the most important components of a successful business. Kaizen is a way to unite everyone among them by sharing the same mindset and approach toward work and development

3. Kaizen increases motivation: Team members are motivated to engage and improve when they see that they are part of the change. When they see their small ideas incorporated into a process of continuous improvement, they are more eager to keep up and contribute.

4. Kaizen helps you to better acceptance of new ideas: When your organization is used to accepting the status quo, new ideas and opportunities can sometimes be seen negatively. With a continuous improvement strategy deployed, your teams will get used to and appreciate working with the notion that change is positive.

The Disadvantages of Kaizen

1. Organizations characterized by limited communication may benefit from prioritizing cultural transformations first to foster a more open and collaborative environment.

2. Kaizen events that prioritize short-term objectives may generate a wave of enthusiasm that does not endure over time and, therefore, fails to yield long-lasting results.

Example of Kaizen

Toyota and Ford are notable examples of companies that have successfully implemented Kaizen principles to improve their operations.

Toyota, in particular, has sustained its success by consistently applying Kaizen. Similarly, Ford has effectively utilized Kaizen to streamline their manufacturing processes and reduce production time. These companies serve as excellent models for the benefits of incorporating Kaizen into organizational strategy.

The Main Tools of Kaizen

Kaizen applies various continuous improvement tools depending on the objective at hand. One example is the 5S tool, which is commonly utilized in Lean Manufacturing to enhance workplace efficiency, productivity, and safety. JIT and Kanban are employed in inventory control, manufacturing, and project management. The five whys method (what, when, where, why, and who) is used to identify the root cause of a problem. Fishbone diagrams and tables are often used for visualizing the problem's root causes. Value stream mapping is an analytical tool that enables the identification of areas to eliminate waste. Follow-up events are tools utilized to maintain and sustain improvements.

In Summary

The secret of Toyota's success story has put the beginning of the Kaizen culture of self-development and eventually continuous improvement. The methodology is easy to implement within every professional or personal scenario, which makes it one of the most famous practices nowadays. So what we learned about it so far?

Regardless of what the popular belief is, "Kairyo" is the literal translation of continuous improvement while Kaizen is better defined as "continuous self-development."

To achieve Kaizen, you need to develop a mindset of self-criticism, reflect on what you have achieved so far and always look for the next highest peak to conquer.

The cycle of kaizen activity: "Plan >> Do >> Check >> Act."

Establishing Kaizen culture is a continuous process. is the external force, but Kaizen is the internal force that drives you to improve regularly.

8. Tools And Technique-Set Up Reduction

Set Up Reduction

A concentrated and structured approach to the reduction or elimination of all activities that take place between the manufacture of the last item and the production of the first good item of a new batch.

Step 1. Set up Reduction & Batch Quantities Cost

Minimum total cost Based on individual machine costs - no link to demand

- Simple trade off – non indication of hidden costs e.g. increased scrap, redundancy, etc.
- EOQ/EBQ is the balance point when total set up and inventory holding costs are minimum.
- Total cost
- Inventory holding cost
- Set up or Ordering cost
- Batch Size

Set Up Reduction

Step 2. Five Step Approach

1. Preliminary Stage
2. Analysis Phase
3. Identify Internal and External Elements
4. Convert Internal to External Where Possible
5. Reduce Internal and External Times

Step3. Identify Internal and External Elements

Internal

Tasks that can only be done whilst the machine/process is stopped.

Examples of Internal Set up

Removing guards.

Removing tools/loading tools.

Removing components.

Adjusting locators etc.

Inspection and test.

External

Tasks that can be done whilst

the machine/process runs on the previous job.

Examples of external Set up

Collecting tools and materials.

Positioning cranes and handling equipment

Preparing tools, clamps, bolts, etc. Clearing up and cleaning.

Planning, organising, preparing data, etc.

Internal (Stopped)

External (Running)

Set Up Reduction Benefits Summary

Reduced set up

- Small Batches
- Avoids waste of overproduction
- Minimises Inventory
- Minimises lead times

Make to Order Philosophy

- Improved production control
- Minimises overheads
- Traditional
- Large Batches / Long Runs
- Encourages overproduction
- Creates buffer inventory
- Long lead times

Make to Forecast Philosophy

- Complex production control
- High overheads

9. Tools And Techniques of Six Sigma

GE Lean Action Workout

Lean Action Work Out (AWO)

Guideline for Lean AWO

Lean Action Work Out is an event in which a Process is identified for analysis and a dedicated team works out a plan for the improvement of the process.

Lean AWO Fundamentals Continuous Improvement Focus (Kaizen) – Focus on Waste Removal using Lean Tools

- Team introduced to Lean Thinking - Focus on Speed
- The key concept is 'Try storming' immediately after Brainstorming - Focus on Action
- New ideas tried quickly and results observed :
- Observe
- Improve
- Repeat
- It is about creativity, not immediate impact on money!
- Suitable for areas where the resolution of a problem has dragged on for a long time
- Suitable when no clarity regarding ownership or

when untrained persons hold key jobs

- AWO – Just Do It! Quick and Efficient Solutions to Reduce Waste! AWO – Just Do It! Quick and Efficient Solutions to Reduce Waste!
- Lean Action Work Out (AWO)
- Guideline for Lean AWO
- Identifies a single process to work on and active facilitation by process owner
- 1-2 days intensive action by a team of Subject Matter Experts of the process area
- Create Current State VSM with the Processing and Wait Time of each step
- Identify Wastes in the processes
- Create Future State VSM and plan to make an impact in a short time (2 weeks)
- Report out to Sponsors and Stakeholders and plan for periodic updates
- Get buy-in from stakeholders for supporting the implementation on time

Lean AWO Activities: AWO – Just Do It! Quick and Efficient Solutions to Reduce Waste! AWO – Just Do It! Quick and Efficient Solutions to Reduce Waste!

Lean Action Work Out (AWO)

Preparation for Lean AWO by Teams

The AWO Event is 'Improve' Phase in DMAIC The AWO Event is 'Improve' Phase in DMAIC

4 Wks Prior →3 Wks Prior→ 2 Wks Prior →1 Wk

Prior→AWO1→dayPresentConductAnalysis,→Assign Time for each→Activity in Process Map,→Refine Scope→Kickoff→Meeting→DefineVOC,CTQ,→Goals & Objectives

LEAN EVENT

Finalize Setup Needs→Finalize Analysis→Continuous Training and Information Sharing/Updates; Team Meetings→Gather Data Define→High-level Process Map,→Set Initial Target Sheet

The AWO Follow-up with Kaizen

10.Tools and Techniques Poka Yoka

Poka-Yoke

Poka-Yoke Another Japanese concept, originated by Shigeo Shingo as part of Toyota Production System. It means, to avoid (yokeru) inadvertent errors (poka). In English, it is termed as "Error Proofing" or "Mistake Proofing". This is a concept by which a process can produce correct results irrespective of the action performed. This is achieved by designing the process in such a way that it is impossible to perform an operation incorrectly. Examples of Mistake Proofing or Poka-Yoke SIM cards are shaped in a way they can't be inserted incorrectly Editors underline syntax errors in red as code is typed Version management processes prohibit developers overwriting each other's code Stringent access control rules prevent WIP code from being copied into Production

Poka-Yoke Mistakes happen in organizations for many reasons, but almost all of them can be prevented, if a standard process exists to identify when problems happen, define root causes, and then take the proper corrective actions. The objective is to prevent, or at least, detect and weed out defects, as early as possible in the process. The use of simple poka-yoke mechanisms can prevent mistakes from becoming catastrophic events. Poka-Yoke eliminates rework on defects and frees time to pursue more creative and value-adding activities. The available classifications of poka-yoke are Prevention Based Poka-Yoke Control Method Warning Method Detection Based Poka-Yoke

Tools and Techniques

Poka-Yoke Prevention based Poka-Yoke Poka -Yoke Sense an abnormality about to happen and signal the occurrence or halt processing based on severity, frequency or downstream consequences Control Method • Senses a problem and stops a line or process for immediate corrective action Warning Method Detection based Poka-Yoke In many situations, it is not possible or economically feasible to prevent defects, (capital cost of the poka-yoke mechanism, far exceeds the cost of prevention). For these situations, defects are detected early in the process, preventing them from flowing to downstream processes and multiplying the cost of non-conformance

• Signals occurrence of a deviation or trend of deviations through a warning

• Avoids serial defect generation.

Example:

Transaction Error

"Please fill all mandatory fields to complete the transaction"

• Does not shut down the process on every occurrence.

• Used when a bandwidth of tolerance exists

• Example: Warnings in C compilation

•Warning: improper pointer/integer combination:

• Warning: old-style declaration or incorrect type for: main If mistakes never happen in the first place, there needs to be no defect fixing! If mistakes never happen in the first place, there needs to be no defect fixing

11. Tools and Technique Quality Tool

7 QUALITY TOOLS

Check sheet

- A simple and effective method of gathering information.
- Ensures consistency of data collected.
- Can be completed whilst doing the normal job.
- Simplifies data collection and analysis.
- Highlights trends.
- Spots problems.

7 QUALITY TOOLS

- Histograms and Measles Charts
- Have much in common with the Pareto Diagram.
- Both show graphically the relative number of occurrences of a range of events.
- Makes important causes become apparent.
- Specification limits can be included to display the capability of the process.
- Can be used to collect data as it happens.
- Histograms may be vertical or horizontal

7 QUALITY TOOLS
Run & Correlation Diagrams

- Used to explore relationships between events and time, and between problems and causes (2 variables).

- Both are experimentation techniques to find out when and how problems arise and how they are rectified.
- Both are simple yet very effective methods.
- Definite trends can be easily detected.
- Pressure

7 QUALITY TOOLS

- Statistical Process Control (SPC)
- SPC is based on tasking measurement of a Process or feature.
- Cause & Effect Analysis
- Identifies possible root causes of current problems.
- Provides a list of areas for which data will have to be collected and then analysed. Having defined the opportunity problem or "effect" in its simplest form, then :
- Assemble appropriate personnel.
- Brainstorm the possible causes.
- Group the causes into broad categories.
- Typically, these might be the 6 M's.
  Analyse the completed diagram and identify the

most likely root cause/s of the effect.

- Measurement
- Method
- Machinery
- Money
- Manpower
- Materials

7 QUALITY TOOLS

- Pareto Analysis
- Simply a frequency distribution of attribute data arranged by category.
- Commonly known as "ABC Analysis" or "the 80/20 Rule".
- Its one of the most effective yet simple tools available.
- An effective ongoing improvement tool.
- Identifies the most significant problems to be worked first.
- Has varying applications for use in manufacturing.
- 80% of the wealth is owned by 20% of the people -80% of holidays are taken during 20% of the year -80% of overtime is worked by 20% of the workers -80% of the work is done by 20% of the people.

12. Tools And Techn. Job Standardization

JOB STANDARDISATION

Job Standardization Why?

Method and time are fundamental to:

- Product costing
- Man-hour/manpower planning
- Capacity planning
- Scheduling
- Performance analysis / review
- Target setting

Benefits

- Enables capable and repeatable processes. Process control at source. Improves accuracy of planning.
- Leads to better adherence to plans. Provides a platform from which continuous improvement can be made.
- Reduces costs.
- Improves quality

13. Tools And Tech- Quality and Service

QUALITY AND SERVICES

Aim

- The quality of the work performed by each employee provides the foundation for the quality of our products and the quality of our sales and service. The combination of these three elements allows Toyota to provide products and services that our customers can use with confidence.

Initiative

- Individual employees involved in each process including development, purchasing, production, sales, and after-sales service, integrate quality into their work. Each function is linked with other processes to maintain the momentum of the quality assurance cycle.

Initiatives Based on the Quality Policy

- Toyota formulates the code of conduct for globally common quality to maintain and enhance the confidence of the customers and discusses a proper response globally and in each region, with the aim of promoting solutions to quality issues and ensuring quality for new businesses and technologies.
- The policy is also shared with affiliated group companies and suppliers to promote collaborative actions for ensuring quality.
- Information about initiatives implemented under the policy is reported to senior management, including board of directors.

Quality Assurance Based on Toyota Quality Control Standards

- Toyota establishes the rules, methods, and criteria necessary for controlling its manufacturing and business processes to enable Toyota to continuously provide the product performance and functions, as well as services, that Toyota aims to achieve.

⬥ Based on the global regulations, Toyota establishes its quality control standards at each production base that are suitable for the customers and environment of each region, and periodically checks and reviews the standards.

Quality Assurance System

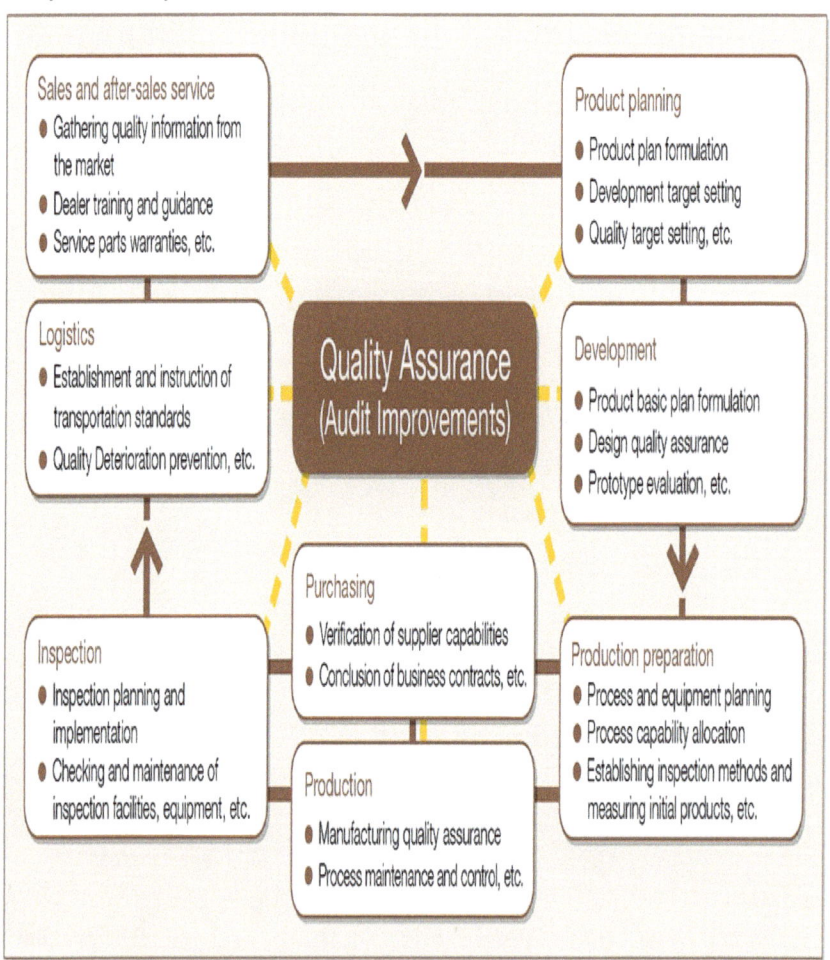

Lean Six Sigma and Action Work Out

References

TCS Global Process Consulting Team Inputs

Capsule Summaries of Key Lean Concepts - Lean Enterprise Institute

Lean Thinking by Joel Gerstenfeld, MIT

C. Fiore; Lean Strategies for Product Development, ASQ, 2003

Lean Sigma: Basic concepts by Helen Bevan, Richard Crowe, Michael O'Connor, Neil Westwood

Institute for Healthcare Improvement (IHI)

www.strategosinc.com/just_in_time.htm

http://www.superfactory.com/OpsConcepts/visualfactory/tabid/97/Default.aspx

http://www.mindtools.com/pages/article/newTMC_5W.htm

http://www.nu-mediadisplays.com/signs/industrial-displays.php

Glossary

Andon	English: 'Sign' or 'Signal'
Genchi	Genbutsu (English: Go and see for yourself): To grasp problems, confirm the facts and analysis of root causes.
Genba or Gemba	The real place, the place where the actual work is done
Heijunka	Production smoothing or leveling
Hansei	Self-reflection
Jidoka	Autonomation — automation with human intelligence
Just-in-time	(JIT) production is a 'pull' system of providing the production sequence – inside and between all the processes.
Konnyaku Stone	The Konnyaku Stone is a small tool used in to smooth body panels before they are painted and to eradicate tiny imperfections.
Kanban	Produce as per customers demand
Kaizen	Pursue Perfection Be tireless rooting to eliminating waste

Kiichiro	Advance Just in Time
Muda	Waste
Muri	Overburden
Mura	Unevenness or irregularity
Nemawashi	Laying the groundwork or foundation; building consensus
Poka-Yoke	mistake proofing
Seirl	Clear out and Classify
Sefton	Configure Put everything in its correct place
SoIso	lean and Check Regular cleaning
Seiketsu	Conformity Good Housekeeping.
Shusuke	Custom and Practice

ABOUT THE AUTHOR

Archana Sharma is a Engineer Tech and Art Crafts and entrepreneur, rebellious marketer, technology expert, and bestselling author.

She has started and grown multiple businesses in various industries, including IT, telecommunications, and marketing.

One of his previous businesses was in the telecommunications in Defence industry, where he faced heated competition from IT Industry, multinational competitors. In four years, Archana grew business from a startup to being Business as one of India's fastest-growing companies, earning.

Archana is passionate about helping businesses find new and innovative ways to leverage technology and marketing to facilitate rapid business growth

www.ingramcontent.com/pod-product-compliance
Lightning Source LLC
Chambersburg PA
CBHW040314220526
45473CB00009B/2424